Preface

We are excited to present the "Science Olympiad All Chapters Combined MCQ Class 2", a comprehensive resource designed to nurture young learners' curiosity and guide them in exploring the wonders of science. This book offers a collection of Multiple-Choice Questions (MCQs) covering all chapters of the Class 2 science syllabus, making it an ideal companion for Science Olympiad examinations.

Science is a journey of discovery, and this book is crafted to simplify learning while fostering critical thinking and problem-solving skills. Each question is tailored to the needs of Class 2 students, ensuring it is both engaging and age-appropriate, while also aligned with the Olympiad exam format.

Key Highlights:

1. **Complete Chapter Coverage**: Thoroughly covers all topics for effective preparation.
2. **Olympiad-Focused Content**: Familiarizes students with the format and level of Science Olympiad exams.
3. **Answers and Explanations**: Provides solutions to enhance understanding and support independent learning.
4. **Skill Building**: Encourages logical reasoning, analytical thinking, and a passion for science.

We hope this book inspires Class 2 students to excel in Science Olympiad competitions and fosters a lifelong love for learning and exploration.

Table of Contents

Chapter 1 : All Chapters Combined MCQs Part 1

Questions

1. Which of the following is a living thing?

a) Rock

b) Tree

c) Water

d) Air

2. What do plants need to make food?

a) Sunlight

b) Moonlight

c) Stars

d) Clouds

3. Which part of the plant grows underground?

a) Leaf

b) Stem

c) Root

d) Flower

4. What is the main source of light on Earth?

a) Moon

b) Sun

c) Stars

d) Fire

5. What do we use to breathe?

a) Ears

b) Eyes

c) Mouth

d) Nose

6. Which of these is not a sense organ?

a) Eye

b) Ear

c) Hand

d) Nose

7. Which animal is known as the 'King of the Jungle'?

a) Tiger

b) Elephant

c) Lion

d) Deer

8. What do we call the young one of a cow?

a) Calf

b) Puppy

c) Kitten

d) Chick

9. Which of the following can fly?

a) Fish

b) Bird

c) Cat

d) Dog

10. What do we get from bees?

a) Milk

b) Honey

c) Wool

d) Meat

11. Which of the following is a nonliving thing?

a) Dog

b) Tree

c) Rock

d) Fish

12. Which of the following grows from a seed?

a) Stone

b) Plant

c) Bird

d) Fish

13. Which part of the plant is colorful and attracts bees?

a) Leaf

b) Stem

c) Root

d) Flower

14. What do we need to stay alive?

a) Books

b) Toys

c) Food

d) Games

15. Which sense organ helps us to see?

a) Ear

b) Eye

c) Nose

d) Tongue

16. Which animal lives in water?

a) Dog

b) Bird

c) Fish

d) Cow

17. What do we call the baby of a cat?

a) Calf

b) Kitten

c) Puppy

d) Chick

18. What do plants give us?

a) Air

b) Water

c) Food

d) Sand

19. Which part of the body pumps blood?

a) Brain

b) Heart

c) Stomach

d) Liver

20. Which of the following is a fruit?

a) Carrot

b) Apple

c) Potato

d) Spinach

21. Which sense organ helps us to hear?

a) Eye

b) Ear

c) Nose

d) Skin

22. What is the main source of energy for plants?

a) Water

b) Soil

c) Sunlight

d) Air

23. Which animal is known for its trunk?

a) Lion

b) Elephant

c) Tiger

d) Horse

24. What do we call the young one of a hen?

a) Calf

b) Puppy

c) Chick

d) Kitten

25. Which of the following can swim?

a) Cat

b) Fish

c) Bird

d) Dog

Chapter 2 : All Chapters Combined MCQs Part 2

26. What do cows give us?

a) Milk

b) Wool

c) Honey

d) Meat

27. Which part of the plant makes food?

a) Flower

b) Leaf

c) Root

d) Stem

28. Which of the following is a herbivore?

a) Lion

b) Tiger

c) Cow

d) Dog

29. What do we use to write?

a) Spoon

b) Pen

c) Book

d) Plate

30. Which of these animals can fly?

a) Elephant

b) Lion

c) Bird

d) Dog

31. Which part of the body helps us to think?

a) Heart

b) Brain

c) Stomach

d) Lungs

32. What do we call the baby of a dog?

a) Calf

b) Kitten

c) Puppy

d) Chick

33. Which of the following is a vegetable?

a) Apple

b) Carrot

c) Banana

d) Orange

34. Which sense organ helps us to taste?

a) Eye

b) Ear

c) Nose

d) Tongue

35. Which animal has a mane?

a) Tiger

b) Elephant

c) Lion

d) Dog

36. What do we get from sheep?

a) Milk

b) Wool

c) Honey

d) Meat

37. Which of the following animals lay eggs?

a) Cow

b) Dog

c) Hen

d) Cat

38. Which of the following is not a source of light?

a) Sun

b) Moon

c) Star

d) Candle

39. Which part of the body helps us to smell?

a) Eye

b) Ear

c) Nose

d) Tongue

40. Which of these is a natural resource?

a) Plastic

b) Water

c) Glass

d) Metal

41. What do we use to cut paper?

a) Spoon

b) Scissors

c) Plate

d) Cup

42. Which animal has spots on its body?

a) Lion

b) Tiger

c) Giraffe

d) Zebra

43. Which of the following is a bird?

a) Cat

b) Dog

c) Sparrow

d) Fish

44. What do we need to drink to stay healthy?

a) Soda

b) Juice

c) Water

d) Tea

45. Which of the following is a farm animal?

a) Lion

b) Tiger

c) Cow

d) Wolf

46. Which part of the plant absorbs water from the soil?

a) Leaf

b) Flower

c) Root

d) Stem

47. Which of the following animals is known for its long neck?

a) Elephant

b) Giraffe

c) Lion

d) Tiger

48. What do we call the baby of a sheep?

a) Calf

b) Puppy

c) Kitten

d) Lamb

49. Which of the following can be used to clean teeth?

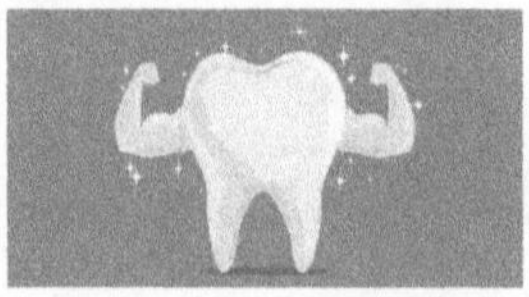

a) Spoon

b) Toothbrush

c) Cup

d) Plate

50. Which sense organ helps us to feel?

a) Eye

b) Ear

c) Nose

d) Skin

Chapter 3 : All Chapters Combined MCQs Part 3

51. What do plants need to grow?

a) Sunlight

b) Darkness

c) Moonlight

d) Stars

52. Which animal is known for its stripes?

a) Lion

b) Elephant

c) Tiger

d) Horse

53. What do bees collect from flowers?

a) Water

b) Nectar

c) Sand

d) Soil

54. Which of the following is a source of energy?

a) Sun

b) Moon

c) Stars

d) Clouds

55. What do we call the young one of a horse?

a) Calf

b) Kitten

c) Puppy

d) Foal

56. Which of the following animals can hop?

a) Dog

b) Cat

c) Rabbit

d) Fish

57. What do plants give out during the day?

a) Carbon dioxide

b) Oxygen

c) Nitrogen

d) Helium

58. Which part of the body helps us to taste food?

a) Eye

b) Ear

c) Nose

d) Tongue

59. Which of the following is not a fruit?

a) Banana

b) Orange

c) Carrot

d) Apple

60. What do we get from trees?

a) Milk

b) Wool

c) Honey

d) Wood

61. Which animal is known for its hump?

a) Horse

b) Elephant

c) Camel

d) Lion

62. Which of these is a source of water?

a) River

b) Mountain

c) Forest

d) Desert

63. What do we call the baby of a goat?

a) Calf

b) Puppy

c) Kid

d) Chick

64. Which of the following animals can crawl?

a) Dog

b) Cat

c) Snake

d) Fish

65. What do we use to brush our hair?

a) Spoon

b) Comb

c) Plate

d) Cup

66. Which of the following is a domestic animal?

a) Lion

b) Tiger

c) Dog

d) Wolf

67. What do plants absorb from the soil?

a) Sand

b) Water

c) Air

d) Stones

68. Which of the following is a reptile?

a) Dog

b) Cat

c) Snake

d) Fish

69. What do birds use to fly?

a) Legs

b) Wings

c) Tail

d) Beak

70. Which part of the plant produces seeds?

a) Leaf

b) Flower

c) Root

d) Stem

71. Which of the following animals has a shell?

a) Dog

b) Fish

c) Turtle

d) Elephant

72. What do we call the baby of a duck?

a) Duckling

b) Puppy

c) Calf

d) Kitten

73. Which sense organ helps us to see?

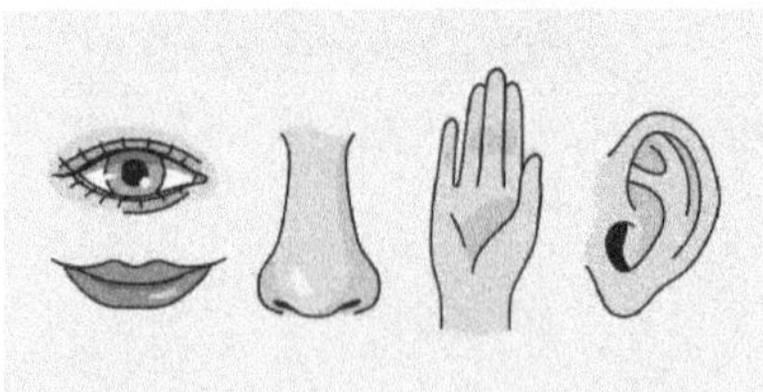

a) Ear

b) Eye

c) Nose

d) Tongue

74. Which animal is known for its long ears?

a) Cat

b) Elephant

c) Rabbit

d) Dog

75. What do plants give out at night?

a) Oxygen

b) Carbon dioxide

c) Nitrogen

d) Hydrogen

Chapter 4 : All Chapters Combined MCQs Part 4

76. Which of the following is a farm animal?

a) Lion

b) Cow

c) Tiger

d) Wolf

77. Which of these animals is known for climbing trees?

a) Fish

b) Dog

c) Monkey

d) Elephant

78. What do we use to see distant objects?

a) Microscope

b) Telescope

c) Binoculars

d) Magnifying glass

79. Which part of the body helps us to hear?

a) Eye

b) Ear

c) Nose

d) Tongue

80. Which of the following is a mammal?

a) Fish

b) Bird

c) Cat

d) Snake

81. What do we call the young one of a lion?

a) Calf

b) Kitten

c) Cub

d) Puppy

82. Which of the following is a source of energy for humans?

a) Food

b) Water

c) Sand

d) Air

83. Which part of the body helps us to walk?

a) Hands

b) Legs

c) Head

d) Stomach

84. What do bees produce?

a) Milk

b) Honey

c) Wool

d) Mea

85. Which animal is known for its speed?

a) Turtle

b) Cheetah

c) Elephant

d) Panda

86. Which of the following is a fruit?

a) Carrot

b) Banana

c) Potato

d) Spinach

87. What do we use to measure time?

a) Ruler

b) Watch

c) Spoon

d) Plate

88. Which part of the plant is used to make clothes?

a) Flower

b) Stem

c) Root

d) Leaf

89. Which of these animals lives in water?

a) Dog

b) Cat

c) Fish

d) Bird

90. What do we get from plants?

a) Air

b) Milk

c) Wool

d) Honey

91. Which of the following is a herbivore?

a) Lion

b) Tiger

c) Cow

d) Dog

92. What do we call the baby of a horse?

a) Calf

b) Kitten

c) Puppy

d) Foal

93. Which sense organ helps us to feel?

a) Eye

b) Ear

c) Nose

d) Skin

94. Which of the following is a natural resource?

a) Plastic

b) Water

c) Glass

d) Metal

95. What do plants need to make food?

a) Sunlight

b) Moonlight

c) Stars

d) Clouds

96. Which animal has a trunk?

a) Lion

b) Elephant

c) Tiger

d) Horse

97. Which part of the body helps us to smell?

a) Eye

b) Ear

c) Nose

d) Tongue

98. What do plants give us?

a) Air

b) Water

c) Food

d) Sand

99. Which animal is known for its spots?

a) Lion

b) Tiger

c) Giraffe

d) Zebra

100. What do cows give us?

a) Milk

b) Wool

c) Honey

d) Meat

Answers

1. Which of the following is a living thing?

b) Tree

2. What do plants need to make food?

a) Sunlight

3. Which part of the plant grows underground?

c) Root

4. What is the main source of light on Earth?

b) Sun

5. What do we use to breathe?

d) Nose

6. Which of these is not a sense organ?

c) Hand

7. Which animal is known as the 'King of the Jungle'?

c) Lion

8. What do we call the young one of a cow?

a) Calf

9. Which of the following can fly?

b) Bird

10. What do we get from bees?

b) Honey

11. Which of the following is a nonliving thing?

c) Rock

12. Which of the following grows from a seed?

b) Plant

13. Which part of the plant is colorful and attracts bees?

d) Flower

14. What do we need to stay alive?

c) Food

15. Which sense organ helps us to see?

b) Eye

16. Which animal lives in water?

c) Fish

17. What do we call the baby of a cat?

b) Kitten

18. What do plants give us?

c) Food

19. Which part of the body pumps blood?

b) Heart

20. Which of the following is a fruit?

b) Apple

21. Which sense organ helps us to hear?

b) Ear

22. What is the main source of energy for plants?

c) Sunlight

23. Which animal is known for its trunk?

b) Elephant

24. What do we call the young one of a hen?

c) Chick

25. Which of the following can swim?

b) Fish

26. What do cows give us?

a) Milk

27. Which part of the plant makes food?

b) Leaf

28. Which of the following is a herbivore?

c) Cow

29. What do we use to write?

b) Pen

30. Which of these animals can fly?

c) Bird

31. Which part of the body helps us to think?

b) Brain

32. What do we call the baby of a dog?

c) Puppy

33. Which of the following is a vegetable?

b) Carrot

34. Which sense organ helps us to taste?

d) Tongue

35. Which animal has a mane?

c) Lion

36. What do we get from sheep?

b) Wool

37. Which of the following animals lay eggs?

c) Hen

38. Which of the following is not a source of light?

b) Moon

39. Which part of the body helps us to smell?

c) Nose

40. Which of these is a natural resource?

b) Water

41. What do we use to cut paper?

b) Scissors

42. Which animal has spots on its body?

c) Giraffe

43. Which of the following is a bird?

c) Sparrow

44. What do we need to drink to stay healthy?

c) Water

45. Which of the following is a farm animal?

c) Cow

46. Which part of the plant absorbs water from the soil?

c) Root

47. Which of the following animals is known for its long neck?

b) Giraffe

48. What do we call the baby of a sheep?

d) Lamb

49. Which of the following can be used to clean teeth?

b) Toothbrush

50. Which sense organ helps us to feel?

d) Skin

51. What do plants need to grow?

a) Sunlight

52. Which animal is known for its stripes?

c) Tiger

53. What do bees collect from flowers?

b) Nectar

54. Which of the following is a source of energy?

a) Sun

55. What do we call the young one of a horse?

d) Foal

56. Which of the following animals can hop?

c) Rabbit

57. What do plants give out during the day?

b) Oxygen

58. Which part of the body helps us to taste food?

d) Tongue

59. Which of the following is not a fruit?

c) Carrot

60. What do we get from trees?

d) Wood

61. Which animal is known for its hump?

c) Camel

62. Which of these is a source of water?

a) River

63. What do we call the baby of a goat?

c) Kid

64. Which of the following animals can crawl?

c) Snake

65. What do we use to brush our hair?

b) Comb

66. Which of the following is a domestic animal?

c) Dog

67. What do plants absorb from the soil?

b) Water

68. Which of the following is a reptile?

c) Snake

69. What do birds use to fly?

b) Wings

70. Which part of the plant produces seeds?

b) Flower

71. Which of the following animals has a shell?

c) Turtle

72. What do we call the baby of a duck?

a) Duckling

73. Which sense organ helps us to see?

b) Eye

74. Which animal is known for its long ears?

c) Rabbit

75. What do plants give out at night?

b) Carbon dioxide

76. Which of the following is a farm animal?

b) Cow

77. Which of these animals is known for climbing trees?

c) Monkey

78. What do we use to see distant objects?

b) Telescope

79. Which part of the body helps us to hear?

b) Ear

80. Which of the following is a mammal?

c) Cat

81. What do we call the young one of a lion?

c) Cub

82. Which of the following is a source of energy for humans?

a) Food

83. Which part of the body helps us to walk?

b) Legs

84. What do bees produce?

b) Honey

85. Which animal is known for its speed?

b) Cheetah

86. Which of the following is a fruit?

b) Banana

87. What do we use to measure time?

b) Watch

88. Which part of the plant is used to make clothes?

b) Stem

89. Which of these animals lives in water?

c) Fish

90. What do we get from plants?

c) Food

91. Which of the following is a herbivore?

c) Cow

92. What do we call the baby of a horse?

d) Foal

93. Which sense organ helps us to feel?

d) Skin

94. Which of the following is a natural resource?

b) Water

95. What do plants need to make food?

a) Sunlight

96. Which animal has a trunk?

b) Elephant

97. Which part of the body helps us to smell?

c) Nose

98. What do plants give us?

c) Food

99. Which animal is known for its spots?

c) Giraffe

100. What do cows give us?

a) Milk

9 789369 721856

Printed at Repro India Ltd.